BEAUTIFUL BIOMES

MARINE BIOME

by Elizabeth Andrews

Cody Koala
An Imprint of Pop!
popbooksonline.com

abdobooks.com
Published by Pop!, a division of ABDO, PO Box 398166, Minneapolis, Minnesota 55439. Copyright ©2022 by Abdo Consulting Group, Inc. International copyrights reserved in all countries. No part of this book may be reproduced in any form without written permission from the publisher. Cody Koala™ is a trademark and logo of Pop!.

Printed in the United States of America, North Mankato, Minnesota

102021
012022   THIS BOOK CONTAINS RECYCLED MATERIALS

Cover Photo: Shutterstock Images
Interior Photos: Shutterstock Images, 1, 5 (top) (bottom left) (bottom center), 6–7, 9, 10–11, 12, 13, 15, 16 (manatee) (mangrove), 19, 20, 21 (top) (bottom)

Editor: Tyler Gieseke
Series Designer: Laura Graphenteen

Library of Congress Control Number: 2021942283
Publisher's Cataloging-in-Publication Data
Names: Andrews, Elizabeth, author.
Title: Marine biome / by Elizabeth Andrews
Description: Minneapolis, Minnesota : Pop!, 2022 | Series: Beautiful biomes | Includes online resources and index.
Identifiers: ISBN 9781098241049 (lib. bdg.) | ISBN 9781098241742 (ebook)
Subjects: LCSH: Ocean--Juvenile literature. | Biotic communities--Juvenile literature. | Habitats--Juvenile literature. | Life zones--Juvenile literature. | Marine animals--Juvenile literature. | Marine plants--Juvenile literature. | Marine ecology--Juvenile literature.
Classification: DDC 577.7--dc23

Hello! My name is
Cody Koala

Pop open this book and you'll find QR codes like this one, loaded with information, so you can learn even more!

Scan this code* and others like it while you read, or visit the website below to make this book pop.

popbooksonline.com/marine-biome

*Scanning QR codes requires a web-enabled smart device with a QR code reader app and a camera.

Table of Contents

Chapter 1
Under the Sea 4

Chapter 2
Three Marine Biomes 8

Chapter 3
Marine Plants 14

Chapter 4
Marine Animals 18

Making Connections 22
Glossary 23
Index 24
Online Resources 24

Chapter 1

Under the Sea

A biome is a large, natural area. It is known for the plants and animals that live there, and its **climate**.

marine

desert

tundra

Watch a video here!

Marine water biomes cover most of the earth. They provide rainwater for land. Plants living in these biomes make lots of oxygen for the planet. Marine biomes come in three types.

Chapter 2

Three Marine Biomes

The ocean is a large body of water. The water is salty. Its temperature ranges greatly all over the world. The ocean near the **equator** is warmer than the ocean near the north and south **poles**.

Some coral reefs are dying. But scientists are trying to find ways to bring them back!

Coral reefs are found in warm, shallow parts of the

ocean. They work as **barriers** along some **continents**.

Estuaries are special. They occur where freshwater rivers or streams flow into the ocean. Some fish and

waterbirds go to estuaries to have babies. These biomes are also safe places for babies to grow.

Chapter 3

Marine Plants

Plants grow in all marine biomes. Plants need sunlight to grow. Some places in the ocean are too deep for light to reach. Plants won't grow there. Near the shore, kelp and algae grow.

Learn more here!

Seagrass meadows and mangrove forests grow in certain marine biomes. They offer food and shelter to many animals.

Manatees and sea turtles eat the seagrass in marine meadows.

Chapter 4

Marine Animals

Many animals live in marine biomes. Coral is sometimes mistaken for a plant. But it is actually an animal. Many grow together to make coral reefs. Starfish and sea turtles live there.

Complete an activity here!

Swimming **mammals** like whales and dolphins live in the ocean. They go to the surface to get the oxygen they need. Fish can breathe oxygen underwater.

The marine biome supports the largest animal on Earth. It is the blue whale. It can weigh as much as 33 elephants!

Birds like herons and ducks make their homes in estuaries. **Crustaceans**, starfish, and clams live in tide pools along some rocky ocean shores.

Making Connections

Text-to-Self

Have you ever seen a marine biome? If you have, which type was it? If you haven't, what type would you most want to see?

Text-to-Text

Have you read a book about any other biomes? If so, how are they similar to and different from marine biomes?

Text-to-World

Some parts of the ocean are struggling because of climate change. Do you know of any ways people are trying to help marine biomes?

Glossary

barrier – a natural feature that blocks or prevents movement, like ocean waves.

climate – weather conditions that are usual in an area over a long period of time.

continents – seven large pieces of land on the globe.

crustacean – any of a group of animals with a hard shell and jointed legs.

equator – an imaginary line around the middle of Earth, halfway between the north and south poles.

mammal – an animal that makes milk to feed its young and usually has hair or fur on its skin.

pole – either end of Earth. The north pole and the south pole are opposite each other.

Index

animals, 4, 12–13, 16–18, 20–21

coral reef, 10, 18

estuary, 12–13, 21

light, 14

oxygen, 7, 20

plants, 4, 7, 14, 16–18

Online Resources

popbooksonline.com

Thanks for reading this Cody Koala book!

Scan this code* and others like it in this book, or visit the website below to make this book pop!

popbooksonline.com/marine-biome

*Scanning QR codes requires a web-enabled smart device with a QR code reader app and a camera.